THE SPC TROUBLESHOOTING GUIDE

THE SPC TROUBLESHOOTING GUIDE

RONALD BLANK

QUALITY RESOURCES.
A Division of The Kraus Organization Limited
New York, New York

Most Quality Resources books are available at quantity discounts when purchased in bulk. For more information contact:

Special Sales Department
Quality Resources
A Division of the Kraus Organization Limited
902 Broadway   800-247-8519
New York, New York 10010   212-979-8600
www.qualityresources.com   E-mail: info@qualityresources.com

Copyright © 1998 Ronald Blank

All rights reserved. No part of this work covered by the copyrights hereon may be reproduced or used in any form or by any means—graphic, electronic, or mechanical, including photocopying, recording, taping, or information storage and retrieval systems—without written permission of the publisher.

Printed in the United States of America

02 01 00 99 98   10 9 8 7 6 5 4 3 2 1

The paper used in this publication meets the minimum requirements of American National Standard for Information Sciences—Permanence of Paper for Printed Library Materials, ANSI Z39.48-1984.

ISBN 0-527-76343-8

Library of Congress Cataloging-in-Publication Data

Blank, Ronald.
    The SPC troubleshooting guide / Ronald Blank.
        p.   cm.
    Includes bibliographical references and index.
    ISBN 0-527-76343-8 (alk. paper)
    1. Process control—Statistical methods. I. Title.
TS156.8.B56   1997
658.5'62—dc21   97-45676
                                                                                  CIP

TABLE OF CONTENTS

List of Figures	vii
List of Tables	vii

INTRODUCTION	
WHAT IS SPC SUPPOSED TO DO?	1
• Process Goals	2
• Product Goals	2

CHAPTER 1	
DOES YOUR SPC SYSTEM NEED TROUBLESHOOTING?	5
• System Level Evaluations	8
• Chart Level Evaluations	10

CHAPTER 2	
WHEN CAPABILITY IS ALWAYS LOW	13
• Reasons for Low Capability	16
• Process Centering	16
• Process Variation	17
• Normality and SPC	18
• Non-normal Distributions	20
• Kurtosis and Skewness	21
• Multimodal Distributions	22
• Multivariate Distributions	23
• Combed Distributions	24
• Edge Peaked and Cut Distributions	24
• Exponential and Weibull Distributions	25

CHAPTER 3	
WHEN THE PROCESS DOES NOT STAY IN CONTROL	29
• Symptoms of Chronically Uncontrolled Processes	31
• Chronically Outside the Control Limits	31
• Chronically Has Runs or Trends	35
• Chronically Has Nonrandom Pattern	37

CHAPTER 4
EFFECTIVE CORRECTIVE ACTIONS **43**
- Why Some Corrective Actions Fail 43
- Corrective Actions for SPC Issues 46
- Corrective Actions for Excessive Variation 49
- Corrective Actions for Chronic Average Issues 50
- Verifying the Effectiveness of Corrective Actions 51
- Conclusion 55

Glossary 57
Recommended Reading 61
Index 63

List of Figures

2-1 Kurtosis
2-2 Skewness
2-3 Multimodal Distribution
2-4 Multivariate Distribution
2-5 Combed Distribution
2-6 Edge Peaked Distribution
2-7 Cut (Truncated) Distribution
2-8 Exponential Distribution
2-9 Weibull Distribution
3-1 Frequently Exceeding One Control Limit
3-2 Frequently Exceeding Both Control Limits
3-3 Incorrect Average
3-4 Nonrandom Pattern
3-5 Run Sum Chart Zones

List of Tables

1-1 Typical Questions for an SPC Audit
1-2 SPC Chart Checklist
4-1 SPC System Symptoms and Causes
4-2 Excessive Variation Symptoms and Causes
4-3 Chronic Average-Related Symptoms and Causes
4-4 Effective Corrective Actions

INTRODUCTION

WHAT IS SPC SUPPOSED TO DO?

Companies that perform statistical process control (SPC) have certain expectations about what the SPC system is going to do for them. When these expectations are unrealistically high, or just plain wrong, the company assumes that SPC does not work. However, it is only when a company properly understands the real goals of SPC and performs SPC properly that it can work well.

An ineffective SPC system may be caused by an organization performing SPC incorrectly or by flaws in the SPC system itself. Symptoms of an ineffective SPC system include consistently poor process capability, charts that are always out of control, and corrective actions that are often ineffective. Troubleshooting an SPC system requires a knowledge of the problems and pitfalls of SPC implementation in addition to their causes and corrective actions.

It is important to understand that SPC does not control processes. *People* control processes. SPC is merely a tool that provides you with the information you need to reduce variation and tell you whether or not your process can meet the customer's expectations. So SPC is really an information gathering method that is used to keep a good process from going out of control. It cannot take an out-of-control process and bring it into control. Only people and a variety of engineering activities can do that. In addition, SPC cannot take a "not capable" process and make it capable.

Process Goals

Statistical process control is applied to a process with two goals in mind. One is to reduce the variation in the process. Control charts help you do this by telling when to adjust the process before it changes significantly or starts producing unacceptable results. The other goal is to improve or maintain the process capability. SPC compares the process average and normal amount of variation to the applicable requirements so you know if the process can meet them.

It does not matter how automated or how labor intensive the process is. It does not even matter whether the process is a manufacturing process or not. The goal of SPC as applied to any process is always the same. Reduce variation and keep the process capable of meeting the customer's expectations.

The reason for trying to reduce variation is that with less variation you have a more predictable and controllable process. This predictability makes for better, more accurate planning. The controllability enables you to reduce waste. Improved predictability and better controllability both result in higher profits.

Product Goals

Just as manufacturing regards improper fabrication or misassembly as a defect in the product, improper performance of a service can be viewed as a defect in the product of service organizations. The goal of companies that apply SPC to their products and services is to prevent defects. Control charts work to prevent defects by alerting people to take action before the process produces a defect.

Another goal of SPC as applied to products and services is to improve productivity. Fewer defects mean less waste. When charting material flow and reaction times, their variation can be reduced. This results in better efficiency, and hence higher productivity.

When these goals are not being met, the SPC system needs some troubleshooting. This book offers ways to diagnose whether an SPC system needs troubleshooting, including guidelines for performing an SPC audit and determining process capability as well as the most effective corrective actions to take as a response to an SPC control chart.

CHAPTER 1

DOES YOUR SPC SYSTEM NEED TROUBLESHOOTING?

When an SPC system is not working properly, the following symptoms may appear:

- The defect rate is highly variable.
- Productivity is highly variable.
- Frequent errors in SPC calculations or procedures occur.
- The cost of doing SPC is unusually high.
- Customer representatives or quality auditors find noncompliances when examining the SPC system.
- Employees complain frequently about having to do SPC.
- The process capability is frequently too low.
- The control charts are frequently out of control.
- Process adjustments are very frequent.

Any one of these symptoms may indicate that there is something wrong with the SPC system. If several of these symptoms occur, however, the SPC system most likely has critical problems that require serious troubleshooting. These symptoms can occur regardless of the type of SPC chart you are using.

Excessive variation in defect rates and productivity are exactly the two things SPC is supposed to prevent. If these symptoms exist in processes controlled by SPC, it is safe to say the SPC system is not working well at all. These symptoms may also indicate that you are applying SPC to the wrong characteristics, especially if the cost of SPC is also high. SPC is most effectively done on process parameters like air pressure, processing temperature, feeding speed, impact force, and heat time. When you apply SPC to product characteristics as an indicator of the process, you must make the measurements as soon as possible after making the part. On a continuously running process, it makes no sense to perform SPC by adjusting the process based on a measurement made several hours after the process produced the part.

SPC errors, quality audit noncompliances, and employee complaints are typically due to inadequate training or training the wrong people. It is not just the machine operators that need SPC training, but also their supervisors, engineers, and managers. Without a proper understanding of SPC, they may provide only lip service to the SPC efforts, and SPC cannot work without their active support.

Low process capabilities, charts always out of control, and frequent process adjustments indicate that the SPC system was incorrectly started or is being carried out incorrectly. They usually happen when a company first begins SPC by doing control charts. An SPC project should start by eliminating the assignable causes of variation *before* beginning to chart the processes. This requires a thorough knowledge of the process and some detective work to identify these causes of variation. After the causes have been determined, you can implement corrective action to eliminate them permanently. Next, check the process to see if it is stable and in statistical control. This can be done by making a pilot run and charting it. If the process stays in control and has no runs, trends, repeating patterns, or out-of-control points after 25 or 30 subgroups, it is stable. Only *after* you achieve stability and have eliminated assignable causes, do you being control charting the actual process.

The person who has the knowledge and authority to adjust the process should be the one to make the SPC measurements. In addition, the adjustments based on these measurements must be made immediately. If the SPC person has to rely on someone else to make the process adjustments, the adjustments will most likely not be immediate. It is inevitable that the person responsible for adjusting the process will have something else to do and not be able to respond to the SPC charts immediately. Consequently, adjustments may occur at the wrong time, if they are made at all.

These symptoms occur due to large-scale or gross problems with your SPC system. They will most likely be obvious. If not, they are typically caught when internal quality audits are performed. Such quality audits must include the auditing of the SPC system. If your company is ISO 9001 or QS-9000 registered, then you are already conducting quality audits internally. Table 1-1 lists typical quality audit questions concerning SPC.

Sometimes, however, problems are more subtle or occur only in specific areas. That is why SPC systems need to be evaluated more closely and more often. Annual or semiannual quality audits will not always detect smaller problems. To identify these more subtle problems, close evaluation of the SPC system should be performed monthly or bimonthly. In addition, this

TABLE 1-1
Typical Questions for an SPC Audit

1. Are managers, engineers, technicians, and operators adequately trained in SPC at levels appropriate for their responsibilities?
2. How do you decide on what characteristics to perform SPC?
3. Is SPC done on all required characteristics?
4. Are written instructions available that indicate on which characteristics to apply SPC?
5. Are written instructions available that tell how to perform SPC?
6. Do written instructions tell the SPC sample size and sampling frequency?
7. Is the type of SPC chart correct for the subgroup size and distribution shape?
8. Are assignable causes of variation addressed?
9. Is the data correctly obtained?
10. Are all of the SPC charts up to date?
11. Are significant process events noted on the chart?
12. What percentage of the SPC charts are not in control?
13. Are process adjustments made immediately when the SPC charts signals that adjustment is needed?
14. How often is process capability checked?
15. What percentage of the SPC characteristics have process capability below the acceptable minimum?
16. Are corrective actions implemented in a timely manner when process capability is too low?

evaluation should occur at two levels, the system level and the chart level.

System Level Evaluations

System level evaluations include periodic reviews of process capability and histograms as well as verifications of training effectiveness. All CpK (process capability with respect to specification closest to process

TABLE 1-2
SPC Chart Checklist

Yes	No	
___	___	1. Is the following information on the chart?
___	___	a) part or process name
___	___	b) part or process description
___	___	c) characteristic being measured
___	___	d) date chart date begins
___	___	e) date chart data ends
___	___	f) required subgroup size
___	___	g) required sample frequency
___	___	h) type of chart
___	___	i) location of data collection
___	___	j) method of data collection
		2. Are the following calculations performed correctly?
___	___	a) process average
___	___	b) control limits
___	___	c) subgroup averages (X-bar charts only)
___	___	d) ranges or moving ranges (R or MR charts only)
___	___	e) average standard deviation (s charts only)
___	___	3. Is the chart up to date?
___	___	4. Is the data plotted correctly?
___	___	5. Is date and time of each data point recorded?
___	___	6. Are process events noted on the chart?
___	___	7. Is there immediate response to out of conditions?

average) values should be checked monthly. It is important to identify which ones are too low and ask for corrective action on them even if the chart shows the process is in control. CpKs may even be plotted on their own chart. If you plot CpKs, you will find they tend to exhibit a certain amount of variation. As long as they do not show an overall downward trend, this variation does not indicate a problem with your SPC system. However, a downward trend in CpK is a definite sign that something is wrong. Histograms should also be periodically checked with current data. Minor to moderate changes in skewness or kurtosis are no cause for concern, but if the type of histogram changes (e.g., from normal to multivariate), it should be corrected. Such corrective actions can prevent problems later.

One way to measure the effectiveness of SPC training is by giving SPC personnel a written test, customized for their position in the corporation and level of training. Another way is to observe them while they are performing SPC. Some companies require an annual SPC recertification test.

Chart Level Evaluations

Chart level evaluations include checking the SPC charts for errors or omissions. Some companies use a checklist to do this. Table 1-2 shows a sample control chart checklist. It was developed for manual control charting but can easily be adapted for your SPC software.

Errors in calculation, plotting, or subgrouping are usually a matter of training. Charts that are not up to date or do not respond when the chart calls for corrective action are a matter of departmental discipline and must be handled authoritatively.

It is sometimes useful to track SPC errors to help determine training needs or improve existing training. A Pareto chart of the types of error being made will reveal what areas need to be emphasized when doing SPC training, and tracking how many errors each SPC person makes can help identify who needs further training.

When SPC is properly started and carried out, the control charts will not be out of control so often. As a result, process adjustments will be less frequent, productivity and defect rates will stabilize, and the cost of doing SPC will decrease. Chapters 2 and 3 of this book help the reader improve process capability and keep the control charts in statistical control. Chapter 4 provides information on solving problems with and implementing corrective actions to improve the SPC system itself.

CHAPTER 2
WHEN CAPABILITY IS ALWAYS LOW

There are different ways of measuring process capability. Which one you should use depends on what it is you want to know about your process. In addition, looking at only one capability index may not be sufficient. The following list summarizes the most common capability indices, explains how to calculate them, and describes what each tells you about your process:

1. Cp (process capability) tells you whether or not the process can meet the specification. Calculate this using the equation:

$$\frac{(USL - LSL)}{6S}$$

where USL = upper specification limit, LSL = lower specification limit and 6S = six times the standard deviation. Cp is also the highest value that either CpK, CpU, or CpL can have when calculated correctly with the process perfectly centered. Improvement to Cp requires elimination of assignable causes of variation and/or better control over normal variation. Occasionally improving Cp requires a major process overhaul.

2. CpU (process capability with respect to upper specification) tells you the capability in relationship to the upper specification limit. Calculate it as:

$$\frac{USL - X}{3S}$$

where X is the process average and 3S is three times the standard deviation. If you multiply this by 3 and use a Z table, it can tell you the probability of producing parts within the upper specification. Subtract this probability from 1 to get the probability of producing a defect exceeding the upper specification. Use this without CpL only when the specification states a maximum without a minimum.

3. CpL (process capability with respect to lower specification) is the process capability in relationship to the lower specification. Calculate this as:

$$\frac{(X - LSL)}{3S}$$

Use it the same way as CpU, but in reference to the lower specification when there is no upper specification limit.

4. CpK (process capability with respect to specification closest to process average) is simply the lesser of CpU and CpL when both upper and lower specification limits exist. On specifications with only one limit, CpK is the same as CpU or CpL as applicable.

5. CpM (process capability with respect to nominal) is the process capability with respect to a

target value (usually the nominal value). Calculate this using the formula:

$$\frac{(USL - LSL)}{6S}$$

where the standard deviation is calculated as deviation from an assigned target value rather than deviation from the average. CpM is especially useful if your nominal value is not the center of the specification. Such would be the case in a specification like 1.500″ with tolerances of +.007 and −.003.

6. Pp and PpK mean the same thing as Cp and CpK, respectively, but they are always calculated from the actual standard deviation (also called RMS deviation). Whereas Cp and CpK are usually calculated using the R/d_2 estimate, Pp and PpK are applied only when the processes are started up for the first time. They are used to test the capability of a new process or to predict what the process capability will be after the process has been running awhile. Cp and CpK are normally applied to ongoing processes for which some history and experience have been developed. Pp and PpK are considered initial process potential. Cp and CpK are considered ongoing process capability.

Which capability index to use depends on what you want to know. Cp tells you if your process *can* stay within the specification, but not whether it *is* staying within the specification. CpU and CpL tell whether or not you are staying within only the upper or lower specification limit, respectively. CpK tells you if you are staying within both of the specification limits. CpM tells you how close you are to nominal.

Reasons for Low Capability

Excessive variation and the process average being off center are the two reasons for low capability. To improve a low process capability, you either reduce the variation or adjust the process average. To tell which will be more effective, you must compare CpK to Cp. If your CpK is significantly less than Cp, then adjust the process average. The greater the difference between CpK and Cp, the more impact adjusting the process average will have. If Cp is low or close to CpK, then you must reduce the variation to improve process capability rather than adjusting the process average. Unsuccessful or too expensive process capability improvement is often the result of adjusting the variation when you should be adjusting the average or vice versa. Both are discussed in the following sections.

Process Centering

Unsuccessful efforts at centering the process may indicate that you are adjusting the wrong element. While adjusting a process parameter can usually move the average, it is not always a process parameter that needs adjustment. The problem could be a tooling issue. When you adjust the dial on a lathe by .002″, how do you know the tool actually moved .002″? In molding operations, a chronically low CpK with a good Cp may indicate the mold itself needs work.

Sometimes the problem with centering a process is caused by personnel or training difficulties. A change in the process center that occurs at the same time as a change in personnel could be due to a difference in technique. Someone may think they are following the prescribed work instructions but may actually be doing something very subtlely different.

Other ways to move the process center include changing to a different lot of material or even changing the location of the process. Do not always assume that adjusting a processing parameter is the correct solution.

Process Variation

If CpK and Cp are close, the remedy is to reduce the process variation. Techniques for doing this are widely published and distributed. Again, do not overlook factors like gage reproducibility and repeatability (R&R), environment, and training when looking for sources of variation. How close to Cp should CpK (or CpU, CpL) be for reducing variation to have a more significant impact than process centering? Here, the 20% rule applies. If your process average differs from the process center by more than 20% of the tolerance, then improve the capability by adjusting the process center. Otherwise improve capability by reducing the process variation. This is a rule of thumb, not a commandment. It has worked well in the author's experience. Another rule of thumb is that if the Cp is more than 1.67, reducing variation will have little impact.

Gage R&R is a frequently overlooked source of variation. The lower your CpK, the more sensitive the process is to gage R&R. When your process average is close to the specification limit, the process spread is more of an issue because it determines how much, if any, portion of the distribution is out of tolerance. A high-gage R&R may add enough variation to exceed the specification limit.

If Cp or CpK is chronically low, you may be charting a part characteristic but adjusting it with the wrong parameter. All of the process parameters that control a particular part characteristic do not have equal influence. Instead, one or two of the several parameters may be the major controlling factors. Designing of experiments (DOE) and other statistical methods can help you decide which parameters are the major influences. The more variables that affect the characteristic, the more difficult it will be to improve and maintain an acceptable process capability.

Normality and SPC

The distribution of the data affects variation and the location of the process center. Not all processes produce normal distributions, and not all design features are normally distributed in nature. Both flatness and surface finish, for example, will not always produce symmetrical curves in nature, and in fact, perfectly normal distributions exist only on computer screens. Some non-normal distributions have no discrete center. Others have a high degree of variation by nature. Studying histograms can help identify ways to improve capability on seemingly stubborn processes.

Many problems in SPC arise because the preliminary work of SPC was not completed. This preliminary work consists of assuring four things:

1. All assignable causes of variation must be eliminated so that only normal random variation is left.
2. The process must be stable (i.e., must be in statistical control, with no trends, shifts in the average, nonrandom patterns, or points beyond the control limits.
3. The process must be capable of meeting the specification with a sufficiently high process capability to satisfy the customer.
4. The person responsible for controlling the process must have a sufficient understanding of

the process and is ideally the same person who plots the chart.

These must be assured *before* you start tracking the process on a control chart. Although a preliminary chart might be used to check stability, do not continue plotting the chart until you have eliminated all assignable causes of variation. You will know your process is stable when you have 25 or more consecutive subgroups within the control limits. A frequency distribution normal enough to pass the skewness and kurtosis tests of normality is a sign that only normal random variation remains.

The formulas for calculating process capability assume you have a normal curve. If your distribution is routinely non-normal despite your best efforts to normalize it, you have to modify the formulas for calculating CpU, CpL, and CpK. For skewed distributions use the mode (highest point on the histogram) instead of the average when calculating capability. For other non-normal distributions, use the median (center of the histogram) instead of the average. Another approach to non-normal distributions is to calculate the process capability as for normal distributions, but calculate the process spread by probability rather than six standard deviations. Consult with an engineer familiar with statistics, or a statistician, for information on this.

In addition to process capability, other SPC calculations need to be modified when the distribution is not normal. When calculating the standard deviation for process capability in non-normal distributions, use the actual standard deviation rather than the Rbar/d_2 estimate. That estimate is valid only for normal distributions. Control limit factors like A_2 and D_3, and D_4 also assume a normal distribution. So if the distribution is not normal, calculate the control limits as the average plus and minus three actual standard deviations, rather than merely estimating them by using these factors. Many SPC software programs allow for using the actual standard deviation for control limit calculation in their setup. Of course, you may use these estimating factors whenever you have a normal distribution.

Non-normal Distributions

If process capability is always too low, check the shape of the distribution. If the distribution is not a normal bell-shaped curve it could result in chronic low capability. The shape of the histogram gives some indication of what is going on with your process. However, sometimes it may not be clear whether a histogram is normally distributed or not. In this case, it is best to avoid making subjective judgments, which are not always accurate. The alternative is to use standard statistical tests that can determine reliably and consistently whether or not a histogram is indeed normally distributed. However, the real question is not whether or not you have a perfectly normal curve, it is whether or not the distribution is normal *enough*. Remember also that normality is an issue only to determine the following:

- Whether to use mode (highest point on histogram), median, or average when calculating process capability.
- Whether or not to use control charts for individual data or X-bar and R charts.
- Whether or not to use actual standard deviation to calculate control limits, or to use control limit factors like A_2.

Always try to normalize the process first by eliminating assignable causes of variation as much as you can because this will result in a more predictable and easily controllable process. Careful study of the histograms can provide valuable knowledge in eliminating sources of variation. The following sections examine several non-normal situations.

Kurtosis and Skewness

Both kurtosis (the relationship of the steepness of the sides of the curve to the end tails of the curve) and skewness (the symmetry of the curve around its average) are indicators of unaccounted-for variables in your process (see Figures 2.1 and 2.2). Experimentation and other statistical techniques can help identify them. If you cannot correct kurtosis and skewness for reasons of economics or technology, then you must compensate for them. Increasing the subgroup size will compensate

Figure 2.1 Kurtosis

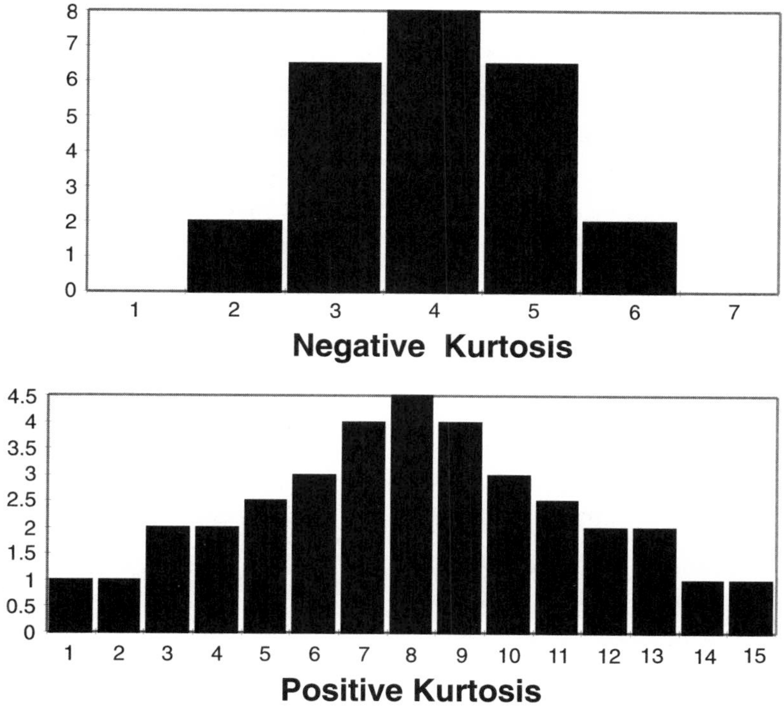

Negative Kurtosis

Positive Kurtosis

Figure 2.2 Skewness

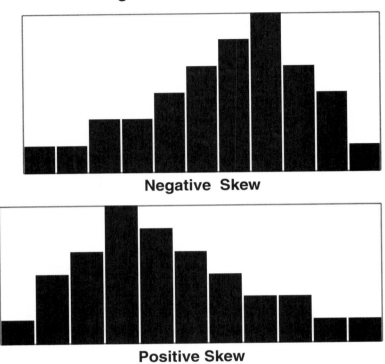

Negative Skew

Positive Skew

for skewness by increasing the effect of the central limit theorem, thereby normalizing your histogram.

Leptokurtosis (steep sides, short tails) actually results in better process capability. Platykurtosis (flattened sides and long tails) is a sign of excessive variation, and you can sometimes compensate for this by increasing the SPC sample frequency.

Although compensating for skewness and kurtosis will improve capability, some compensation will do more for an out-of-control situation than it will for process capability. Your best bet is to try to find the cause and eliminate it.

Multimodal Distributions

Multimodal distributions, like the one shown in Figure 2.3, indicate your measurements are coming from dif-

Figure 2.3 Multimodal Distribution

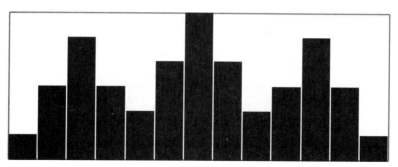

ferent sources (e.g., different mold cavities, different machine fixtures, or different material lots). The modes are the high points in the curve. Two modes indicate two different sources. Three modes indicate three sources, and so on. Curves with more than one mode will have high standard deviations, and therefore low capability. If you cannot make the sources more alike, then treat them separately with each source having its own SPC chart. Individually, their capabilities may be acceptable.

Multivariate Distributions

Multivariate distributions indicate there are many variables at work causing excessive variation. An example is shown in Figure 2.4. This usually happens when assignable causes were not addressed or eliminated and the process was not stabilized before control charting

Figure 2.4 Multivariate Distribution

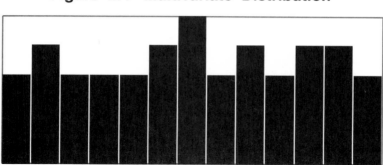

Figure 2.5 Combed Distribution

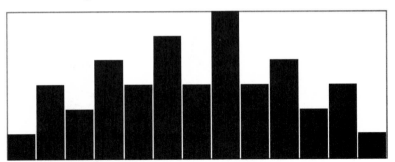

began. The amount of variation in a mix of variables like this will cause a high standard deviation and therefore a low process capability.

Combed Distributions

Resembling the teeth on a hair comb (see Figure 2.5), this is an easy problem to correct. The cause is either excessive rounding off of measurements, or not measuring with enough resolution (i.e., using the smallest increment the tool can measure). If your distribution always looks like this, make sure no one is rounding off the measurements excessively. In fact, as a rule, avoid rounding off measurement results altogether in SPC. When you must round off, do it to only one-fourth of the measuring device resolution and never beyond the number of decimal places in the specification. Better yet, change to a different measurement tool or method so rounding off is not necessary. The resolution of your measurement method should ideally be no more than one-tenth of the increment on the specification. Rounding off data is itself a source of variation and can result in lower CpK.

Edge Peaked and Cut Distributions

Edge peaked and cut (or truncated) distributions are alike in that they both look as if one side of the distribution has been cut off with a blade. Figures 2.6. and 2.7

Figure 2.6 Edge Peaked Distribution

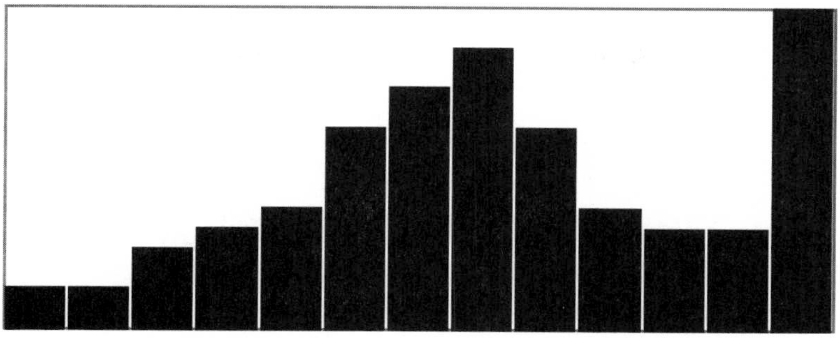

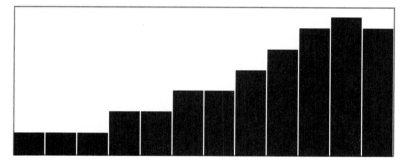

Figure 2.7 Cut (Truncated) Distribution

illustrate these conditions. The "cut" is sometimes the result of some natural limit of your material, process, or equipment setting such as temperature or pressure. It may even be a natural limit of what you are measuring, as flatness has a natural limit of zero. More often it is the result of a sorting operation, whether intentional or not. An operator who picks and chooses samples carefully may be unintentionally sorting out variation. In the case of an edge peaked distribution, the peak at the edge may be due to adjustments or a self-correcting process. Some highly automated assembly machines inspect and test product, then automatically adjust the process or the product when out of specification.

Exponential and Weibull Distributions

An extreme case of skewness may appear to be an exponential distribution. Curves with positive skew will look like a Weibull distribution (see Figures 2.8 and 2.9). Do not be fooled by these appearances. If the data was obtained from an ongoing stable process, then consider them skewed distributions. True exponential and Weibull distributions are of operating lifetimes and wear out, and so are used primarily in reliability. Conventional SPC is intended for ongoing processes.

For attribute SPC where *p, np, c,* or *u* charts are used, you still need to stabilize the process and come as close to the shape of a normal curve as you can. These charts are associated with attribute or discrete distributions, namely the binomial, Poisson, or hypergeometric. These distribution curves all closely approximate the normal bell curve in shape. Therefore, if

Figure 2.8 Exponential Distribution

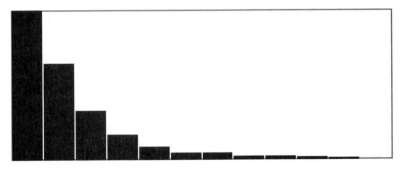

Figure 2.9 Weibull Distribution

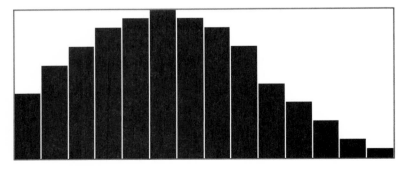

you make a histogram of the attribute SPC data, you should still see the familiar bell shape. If you do not, then there is something affecting either your process or your measurements. If you cannot come close to the normal shape, then consider using a variables type chart when possible.

CHAPTER 3

WHEN THE PROCESS DOES NOT STAY IN CONTROL

SPC charts tell us when to make adjustments to the process. The more stable the process is, the less often adjustments should be necessary. A chart that calls for adjustment too often indicates something is wrong. But what is too often? Whether or not out-of-control situations occur too frequently is something only you can judge for yourself. Your judgment will be valid, however, only when you have eliminated assignable causes before charting and are applying the right rules to the right type of chart. You must also properly determine subgroup size and frequency.

There are a few universal SPC principles that when violated will result in your SPC chart being out of control very frequently. Check that you are not violating any of these principles before you search for other reasons why your chart is chronically out of control. Principles 1 through 4 apply to attribute type control charts as well as to variables type charts. Violation of one or more of these five principles is the most common reason for a control chart constantly indicating an out-of-control situation:

1. The use of control charts should not begin until *after* you have removed all assignable causes of variation and the process has stabilized. Removing assignable causes of variation and stabilizing the process is the first step in setting up an SPC system. Do this before you apply the control limits to production.

2. The rules by which you interpret control charts are not universal. The so-called Western Electric Rules are applicable only when you have a normal distribution. They are not used by all companies in all industries.

3. You must use the right kind of chart for the situation. Charts for individuals and moving ranges (called Ix-MR or X-MR charts) are applicable only when you have a normal distribution. The shape of the frequency distribution does not matter when you are using an X-bar and R chart. If you are using attribute type SPC charts, you should know that p charts assume a binomial distribution while c charts and u charts assume a Poisson distribution. Both of these will closely approximate the normal bell shape but will not meet it exactly. Consequently, they may fail a test of normality even when there is no assignable cause.

4. Subgroup size and sampling frequency are not arbitrary. They are determined by the distribution shape. Subgroup size of one to three is sufficient for a normal distribution. As the distribution shape departs from normal, you need to increase the subgroup size. For highly skewed distributions, a subgroup size of eight or more may be necessary. Always try to normalize your data by removing assignable causes from

the process first. Increasing the subgroup size should be a last resort.

5. Control limits are often calculated by multiplying the average range by a factor such as A_2, A_3, or E_2. This produces valid estimates of three standard deviations. However, these factors all assume a normal distribution. If your distribution is not normal, you must calculate the control limits by multiplying the actual standard deviation by three.

Symptoms of Chronically Uncontrolled Processes

With these facts in mind, we will now examine the symptoms and causes of SPC charts that are chronically out of control. There are basically three chronic out-of-control situations:

1. Data points on the chart are outside the control limits too often.
2. The chart frequently shows runs or trends.
3. Nonrandom patterns continue in spite of corrective actions.

All of these conditions can be caused by not applying any one or more of the five SPC principles. Each of these conditions is examined in the following sections.

Chronically Outside the Control Limits

If in your judgment the number of points that are outside the control limits is always too frequent, it is usually an error in SPC procedure. If you are using *p, c,* or *np* charts, verify that the calculations for control limits are correct. When using *c* or *np* charts, the person doing the counting of defects must understand the difference between counting the number of defects and the number of defective parts. If counting only the number of defective parts, not the number of defects each piece has, then the chart should be an *np* chart. If counting the defects themselves, rather than the number of defective parts, then use a *c* chart.

If the SPC is being done manually, then the most common mistakes are mathematical. Start by checking if the averages, ranges, and control limits are correctly calculated. Then determine if the control limits and data points are correctly plotted.

When calculations are correct, or if you are using SPC software, the errors may be in the way you practice SPC. Examine the histogram. What kind of distribution do you have? Refer to Chapter 1 if necessary. Is the kind of chart you are using correct for the kind of distribution you have? Remember that charts for individual and moving ranges require a normal distribution, while X-bar and R charts can be used for any distribution. If the X or X-bar chart stays in control, but the R or MR chart is frequently out of control, you most likely have a non-normal distribution. If the subgroup size is more than ten, you may want to use an X-bar and S chart rather than X-bar and R. You must use X-bar and S charts if the subgroup size is more than 15.

Positive kurtosis may show on the control charts as frequent out-of-control points exceeding both the upper and lower control limits as shown in Figure 3.1. If your histogram has this condition (see Chapter 1), increase the SPC sampling frequency. If your R or MR chart is frequently out of control, fix your SPC interval by the number of pieces produced, rather than as a time interval (e.g., every 50 pieces instead of every two hours).

Some control charts frequently exceed the same control limit, either the upper or the lower, but not both (as shown in Figure 3.2). This may indicate that your

Figure 3.1 Frequently Exceeding Both Control Limits

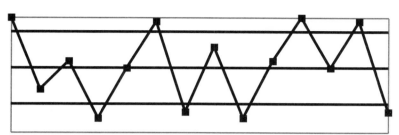

histogram is skewed. If this is the case, there is an assignable cause of variation at work on your process. Identify and correct that cause. Only if you cannot correct it should you then use an X-bar and R chart to solve the problem. If you are already using an X-bar and R chart and your data is always skewed or the distribution appears cut, you may need to increase the subgroup sample size. Typically the subgroup size should be more than five if your distribution is always skewed. If you are using SPC software that gives you a skew calculation, then multiply that by two and round off to the next highest number. Use that as your subgroup size.

Frequently exceeding one control limit and not the other can also indicate that a long-term change in the average has taken place since the last time the limits were recalculated. If this change in average is deliberate, or you decide to keep the average where it is, then

Figure 3.2 Frequently Exceeding One Control Limit

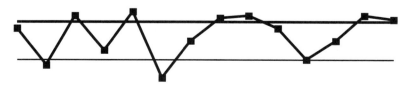

recalculate the control limits. Otherwise, apply a corrective action that will bring the average back to the center of the control limits.

A combed distribution may show out-of-control points frequently due to rounding off of the data. Rounding up may cause a measurement to exceed the upper control limit that it otherwise would still be within, had rounding off not occurred. Similarly, rounding down may cause a measurement to be below the lower control limit when it really is not.

If rounding off data or your distribution shape is not the issue, there are several other reasons a control chart may be chronically out of control. Has the process been modified since the last time the control limits were calculated? Every time you make a deliberate change to the process you should collect data and recalculate the control limits. You want 20 to 30 data points to be included in your control limit calculation. Many companies do a separate process capability study after each permanent deliberate process change.

Do not overlook people and measurement technique. Is there more than one person making these measurements? If so, are they measuring the same thing in exactly the same place, in the exact same way? How repeatable is the measuring equipment?

Overcontrol is another problem. Some operators think they are reducing variation by adjusting the process toward nominal even when the chart does not call for it. Sometimes adjusting at a fixed interval, other times adjusting after every SPC measurement! Such constant adjustment actually increases variation and makes it harder to stay in control. Do not adjust the process unless the control chart indicates it actually needs adjustment.

One final point to realize is that a control chart is really only a tool to identify when to investigate a situation. Charts do not control processes. People control by making decisions and acting on them. Sometimes people choose to do certain corrective actions—rather than the actions that might be indicated by the results of an investigation—out of habit. Other times they are merely doing what they are trained to do, whether or not it is the proper corrective action. Always consider an out-of-control situation to be a call for investigation. It is not an order to take action.

Sometimes the investigation leads to the conclusion that no action should be taken. For example: You have a run of eight points below the process average and the specification has a maximum only (such as flatness or surface finish). This lowering of the average then represents an increase in CpK, thus better quality and fewer defects. You investigate why it happened so you know what it takes to improve your process, but why should you deliberately lower your CpK and increase the defect rate by adjusting the process average back up again? Instead, allow the process average to change and recalculate the limits to fit the new average. Some might think this is "the tail wagging the dog," but it is actually a learning opportunity. The chart triggers you to investigate why it happened so you can understand your process better and improve it when you want.

Chronically Has Runs or Trends

Ideally, there should be approximately 50% of the points on each side of the average. An occasional majority on one side is not cause for concern, but an obvious majority always on one side is cause to investigate. When a control chart frequently shows runs of seven or more points in a row all on the same side of the average, it indicates the actual process average has changed and the chart therefore has the incorrect average (see Figure 3.3). This condition can occur on both

36   THE SPC TROUBLESHOOTING GUIDE

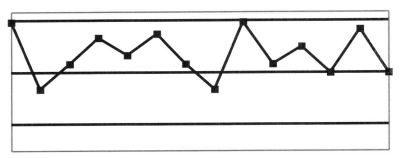

Figure 3.3   Incorrect Average

attribute and variables type charts. The corrective actions necessary are the same. This change may or may not be deliberate.

A majority of points constantly on one side of the average, with or without distinct runs, also indicates a change in the average. If your attempts at adjusting the process average back to where it was are unsuccessful, it may be that the parameter you are adjusting is not acting alone. Such would be the case if two or more process parameters interact with each other. Statistical techniques like design of experiments (DOE), regression analysis, and analysis of variance (ANOVA) can all help identify and understand the effects of interactions. See the list of suggested readings in the back of this book. Consult with an engineer familiar with statistical methods.

You may also try collecting enough data to construct a new histogram. Examine the new histogram and take action appropriate to the kind of histogram you have, as indicated in chapter 1. Then recalculate the process capability and take such actions as necessary. Finally, recalculate the control limits. Note that every time the process is purposely modified you must create a new histogram using data only from after the modification. Then recalculate the process capability and control limits using the same data as for the histogram. Always recalculate the control limits and process capability after you have made a deliberate change to the process and have collected enough data.

Do not confuse process adjustments with deliberate process changes. A process adjustment merely adjusts the value of some processing variable while the process is running in order to make the process behave the way it did before the adjustment was necessary. Process changes are planned, intentional, more or less permanent, engineering changes to the way the process is designed and implemented. These usually involve some downtime. They may even involve the addition or removal of one or more process steps. Whereas process adjustments are often short lived, process changes are intended to be permanent.

When the value of the SPC measurement is consistently decreasing or increasing, it is showing a downward or upward trend. This too is a change in the average, but it is a gradual change. Therefore, it has a gradual cause such as tool wear, warming of the machine, operator fatigue, or anything else that comes on gradually. It may be so gradual as to be seasonal. Remember that sudden changes usually have sudden causes and gradual changes have gradual causes. Do not overlook such gradual causes as gauges going out of calibration, room temperature, material age, employee health, or working conditions.

Chronically Has Nonrandom Pattern

Broadly defined, a nonrandom pattern is one that is produced by anything other than normal random variation. Simply put, it is produced by an assignable cause. By this definition, any undesired condition can be considered a nonrandom pattern, including constant instability with many out-of-control points, gradual trends, and shifts in the average. These have been discussed previously; this section discusses cyclic patterns and run sum violations as causes of nonrandom patterns.

Cyclic Patterns

Any pattern that repeats over and over again on a control chart has a cause that repeats as well. Repeating patterns on a chart, as shown in Figure 3.4, indicate that the measurement being made is sensitive to something that happens at fixed intervals. However, this does not necessarily indicate that something is wrong with the process. All manufacturing processes that produce quantities of the same product are really repeating sequences of planned events. Preventive maintenance, adding a new lot of material when the old lot runs out, and changing operators at the end of their work shift, are all normal parts of any manufacturing process. If they occur at fixed intervals, they may show on control charts as a cyclic, repeating pattern, but they are not necessarily problems.

Remember that the control chart is only a tool to tell when to investigate. As long as the cause is known to be a routine necessary occurrence, no corrective action is required. You should note the cause on the control chart. If these normal routine activities cause a significant probability of producing defective results, then it is wise to make the process less sensitive to them.

Figure 3.4 Nonrandom Pattern

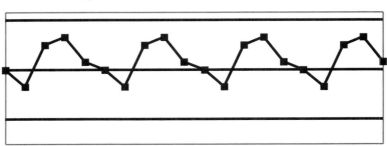

If normal, routine events during processing are not occurring in step with the cyclic pattern you see on your chart, then there is an assignable cause at work. It can be the result of a variable that had not been considered before. It can also be caused by some*one* rather than some*thing*. People are truly quite habitual in their work. Someone may have a work habit that affects the process or the measurement in such a way as to show on the control chart. This is most likely the case if you see a spike occurring at a more or less fixed interval. Cyclic behavior is also common with operations occurring on more than one shift. Observe work habits carefully and over a long enough period to find a human cause. This kind of cyclic pattern can be prevented by having all operator training done by the same person and not relying on one shift to train the other.

Many times it is the measurement results that are changing at a fixed interval, rather than the process. For example, a slot is measured to five decimal places using a coordinate measuring machine (CMM) every two hours, but every day at 2 PM someone else is using the CMM. At that time the slot is measured to only three decimal places on a dial caliper instead. This difference in measurement method causes a blip in the chart at the same time every day. This is a cyclic nonrandom pattern caused not by the process but by the measuring system. Therefore, process adjustment is not needed. What is needed is a standardized way of measuring and sufficient equipment to do so.

Run Sum Violations

Another type of nonrandom pattern is the run sum violation. There are rules that specify what proportion of the points on a chart can be located where in relation to the center line. These are called run sum rules, and they vary from company to company. It is not unusual for them to be dictated by the customer. The control chart is first divided into zones. Zone A is ± three standard deviations from the center line. Zone B is ± two standard deviations from the center line, and Zone C is one standard deviation from the center line (see Figure 3.5).

Examples of run sum rules are that if two of three points are in zone A or four out of five are in zone B, it is a call to investigate. The problem with run sum rules is that they all assume you have a normal distribution. Some companies apply them to both attribute and variables charts. What they actually indicate is that your histogram has lost its normal bell shape, which indicates an assignable cause at work if the data was normally distributed beforehand. But not all measurements are normally distributed in nature. A histogram that is naturally not normal will constantly violate run sum rules even when assignable causes have been eliminated. They should not be applied to such situations. If you are required to use certain particular run sum rules, remember that they are valid only when you have a normal curve and they indicate only that you have lost normality. If you had normality to start with, then by all means respond to them. A loss of normality does indicate an assignable cause at work. But if your distribution shape was not normal to begin with, these run sum rules will be constantly crying out for needless investigation.

If there is a run sum violation and it cannot be attributed to any assignable cause, then you can increase the subgroup size to make the central limit theorem more effective.

As stated, some characteristics are not normally distributed in nature. If this is truly the case, then run sum rules are not applicable. Other times processes have variables acting on them that are beyond your control. You can deal with this when using run sum rules by

Figure 3.5 Run Sum Chart Zones

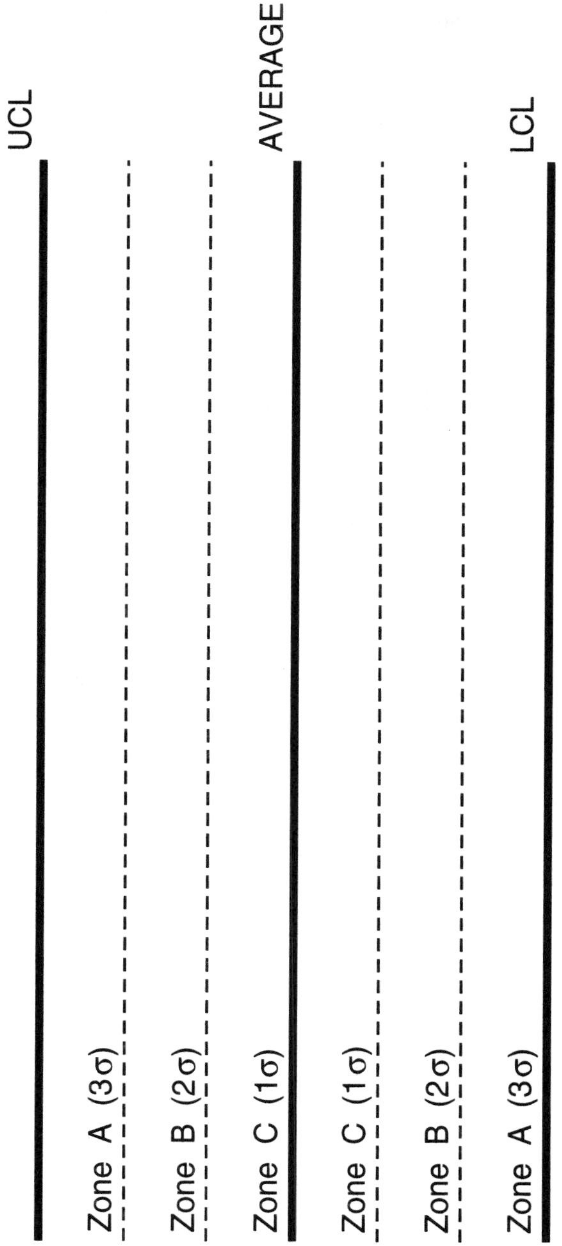

UCL

Zone A (3σ)

Zone B (2σ)

Zone C (1σ)

AVERAGE

Zone C (1σ)

Zone B (2σ)

Zone A (3σ)

LCL

increasing the subgroup size. Subgroups cause a normal bell shape by applying the central limit theorem. Simply stated, this theorem says that when you calculate averages from a group of samples from the same population, the averages will be normally distributed. Increases in subgroup size need not be long term. If the cause of the non-normal situation has been corrected, then you can return to the smaller subgroup size.

Normality is one reason for subgrouping in X-bar and R charts. You perform SPC on the averages of the subgroup samples. The central limit theorem makes these averages normally distributed. Increasing the subgroup size increases the effect of the central limit theorem. This causes the normality to show itself with less data and be much more resistant to changes. So if you must use run sum rules and your histogram never has had a normal bell shape, then increase the subgroup size. Use caution here. Increasing the subgroup size really only hides the lack of normality. It does not cure it. Besides that, increasing the subgroup size is expensive in time and materials. It should be used only as a last resort and should not be taken lightly. Try everything you can to normalize the process first.

The next chapter discusses why some corrective actions are ineffective and matches corrective actions to stubborn chronic control problems. Also in the next chapter are methods of verifying the effectiveness of corrective actions.

CHAPTER 4
EFFECTIVE CORRECTIVE ACTIONS

Why Some Corrective Actions Fail

Corrective actions taken as a response to an SPC control chart can fail for a variety of reasons, including:

- They are not completely implemented.
- They are not correctly implemented.
- They are not sufficient to affect the process.
- They address the wrong process variable.
- The variable adjusted interacts with some other variable that was not adjusted.
- Their effect is only temporary.

When implementing any corrective action, it must be done completely and correctly to have the intended effect. This may seem obvious, but it is worth noting because many times we believe we are doing something correctly and completely, when actually we are not. A complete corrective action addresses the root cause of a problem *and* prevents it from happening again. Training and standardized, documented procedures can prevent implementations that are incomplete, incorrect, insufficient, or of only short-term effect.

When you adjust something in response to a control chart, the adjustment must be of sufficient magnitude to have the desired effect. It must also be the right adjustment to make. If the effect is only temporary, investigate why, then determine the time it will last and then adjust again at the necessary intervals.

Corrective actions in SPC are usually just a process adjustment, but sometimes a full-blown engineering type corrective action is necessary. When this is the case, taking corrective action must include four steps to be truly effective:

1. You must completely describe the problem. Tell the technical representative who, what, where, when, why, how, and how many (or how much). The more completely and specifically you can describe the problem, the more the technical personnel can zero in on precisely what action to take.

2. There must be a corrective action to contain the problem. This is usually called the containment or short-term corrective action. Such an action must prevent the spread of the problem and limit the amount of production affected. It must also shut off the symptoms to the extent that they are not visible to the customer.

3. There must also be a permanent corrective action that completely eliminates all of the root causes.

4. There must be a separate preventive action that addresses the nonphysical, non-engineering situations that allowed the root causes to exist in the first place. This includes such factors as company policies, training practices, cultural attitudes, and any human behavioral factor that allowed the root causes to exist prior to the surfacing of the problem.

Again, these principles would be included in corrective actions that are not mere process adjustments. Such corrective actions might be done early in SPC, for example, when eliminating assignable causes of variation prior to control charting or when attempting to make a noncapable process into a capable one. You obviously would not apply all of the steps simply to restore a previously controlled process back into control. However, you may need to apply all of them to fix a problem with the SPC system itself.

Any corrective action involves change. Resistance to changes associated with a corrective action is natural and common, but it can substantially reduce the effectiveness of the corrective action. It is therefore essential that you investigate the potential for resistance and then present the changes in a way that will alleviate people's concerns before the actual changes take place.

For example, one reason people might resist the changes associated with a corrective action is that they may be unsure about their ability to do whatever the change requires of them. Others may resist a change because they don't view it as a necessary change. Thorough training customized for the individual can be a great help in gaining the cooperation and support necessary for a corrective action to be fully effective.

Prejudices against a person, tool, or method can also cause resistance to change and reduce the effectiveness of a corrective action. These prejudices can be overcome if the person resisting the change is involved in the development of the corrective action. Often it is a supervisor or operator that resists the change. Because operators are intimately involved with the process, and supervisors usually have considerable knowledge and experience, they can actually be valuable resources in developing the corrective action.

Many companies now use a team approach to problem solving, and supervisors and operators can be valuable members of such teams. Individuals who are involved in the development of the corrective action will typically offer much less resistance to change when it is time to implement it.

Corrective Actions for SPC Issues

One factor to be considered when determining the need for a corrective action is the frequency with which the problem occurs. A trend can be caused by tool wear, but if that tool is wearing out every week, it indicates that the tool simply is not up to its demands and should perhaps be made of a more durable material. Similarly, when developing corrective actions, you must consider the frequency with which adjustments are necessary. If the corrective action is determined to be to continually open or close a valve, you should consider how often you'll need to do this. Does the frequency of adjustment indicate the valve is unstable? If so, you may need to replace the valve rather than simply adjust it. If you are making the same adjustment over and over again, you need to examine why and determine what you can do about it.

Another factor to be considered as part of determining the need for corrective action is whether or not to adjust the average or the amount of variation. As mentioned earlier, the relationship between Cp and CpK can help you determine whether or not to do this.

You must also consider interactions between process parameters when implementing corrective actions in SPC. Plastic molding, ultrasonic welding, and fully automated assembly operations involve many adjustments that interact strongly with each other. Adjusting one of these aspects of a manufacturing system very likely might cause the need to adjust another or several of them. Bringing one into control may push another out of control. A good understanding of your process is essential when taking corrective actions in an environment with such interactions.

In addition to being taken in response to out-of-control conditions on control charts, corrective actions must also sometimes be taken in response to problems that occur in the SPC system itself. These problems are not related to any particular process that you are trying to control and are not corrected by process adjustments, but they require corrective action just the same. Their symptoms indicate a problem with the way you are doing SPC or the way the SPC system is designed. Table 4.1 lists SPC system issues by symptom and gives their causes.

These problems can sometimes be the result of inadequate training, or training that emphasized the wrong issues. Many SPC training programs emphasize histograms and control chart construction. Both are important to SPC, but they are not how to begin SPC. You begin by finding and eliminating the assignable causes of variation in your processes. Many SPC training programs do not place enough emphasis on this. Consequently, many companies begin SPC by charting the process right away. They expect the charts to get their process into control. This simply does not happen. SPC charts only tell you when the process is going out of control, but it must be in control first. That requires removing assignable causes of variation from your process. Do this first, then begin charting.

When training people for SPC, you need to emphasize different aspects of SPC according to whom you are training. When training operators, emphasize chart construction and responding. When training engineers, emphasize process capability and assignable cause removal. On the other hand, everyone should be taught the theoretical basis for SPC because this gives the system credibility.

Other common causes of SPC system problems are listed in Table 4-1.

TABLE 4-1
SPC System Symptoms and Causes

SYMPTOMS	CAUSES
Frequent math errors or incorrect control limits	Insufficient SPC training Insufficient time to do SPC Insufficient data for valid calculations
Lack of response to SPC charts	Lack of chart visibility on production floor Insufficient understanding of the process being controlled Appropriate people not trained Appropriate people not doing SPC Person plotting the chart is not the same person responsible for process adjustment
Too many SPC charts to handle	Method of choosing what to do SPC on is incorrect Doing SPC on product characteristics rather than process variables
SPC policies and practice vary between departments	SPC not standardized with written policies and procedures All SPC personnel not trained by the same person or in the same way

Problems with the SPC system can be prevented from recurring by not repeating the mistakes that caused them in the first place and by verifying that the corrective actions are correctly and completely implemented.

Periodic evaluation of the SPC charts, process capabilities, and personnel can help prevent smaller, more subtle problems from occurring. Such evaluations can take the form of monthly audits and need not cover all processes each time.

Corrective Actions for Excessive Variation

When your control chart has too many points in zone A, or frequently goes out of the control limits on both sides of the average, it indicates that you need corrective action for excessive variation. These are often the most difficult corrective actions to identify. Address them as best you can. If you absolutely cannot eliminate or correct the cause, then you have to make your process less sensitive to causes of variation. First, identify the causes of variation such as material, human factors, measurement, or environment. Then make your process less sensitive to them. One way to do this is to isolate the process from the cause. If your process is sensitive to environmental changes like temperature or humidity, try insulating the process from these changes by applying temperature insulation adding a humidity control device. If your measurements are highly influenced by the person doing the measuring or have poor repeatability, use a different measurement method. If material lot differences are a constant source of variation, try planning the work so that frequent material lot changes are not necessary.

Determining corrective actions for excessive variation is easier when you know what is causing the variation. Table 4-2 lists symptoms of excessive variation and their most common causes.

TABLE 4-2
Excessive Variation Symptoms and Causes

SYMPTOMS	CAUSES
Frequently out of both control limits	Limits calculated incorrectly Poor gage R&R Assignable causes of variation not removed before starting control charts Frequency distribution has high kurtosis Insufficient understanding of the process being controlled
Variety of run sum violations occurring too frequently	Non-normal frequency distribution Assignable causes of variation not removed prior to charting Poor gage R&R

Sometimes more advanced statistical techniques are necessary to identify causes of variation and eliminate them. This may require doing DOE, ANOVA, regression analysis, or another problem-solving technique. Consult with a quality engineer or statistician for help, and see the list of suggested readings at the end of this book.

Corrective Actions for Chronic Average Issues

Taking corrective action for the average, rather than correcting for excessive variation, is necessary when more data points are on one side of the average than on the other. It is also needed when the average is close enough to the specification limit so that CpK is low. Corrective action for the average is also needed when a sudden shift in the average occurs. This is usually just a matter of adjusting some process parameter like temperature or air pressure or even a matter of adjusting a cutting tool. If your adjustments are not effective, you may be adjusting the wrong thing, or there may be a strong interaction with another variable so that you have to adjust more than one. You could also simply not be adjusting enough to have the desired effect. If you are not sure what to adjust, ANOVA can steer you to what has the most variation, which is usually the home of the culprit. DOE can rank your variables in order of degree of influence, thereby telling you what to adjust. Regression experiments can help identify cause-and-effect relationships. Consult with someone with a knowledge of industrial statistics. Table 4-3 lists symptoms and causes to help you determine corrective action on SPC issues dealing with the process average.

As you can see, various symptoms have some causes in common. Other causes are unique to specific symptoms. Table 4-4 lists common corrective actions for the causes identified in Tables 4-1, 4-2, and 4-3. These are proven effective in a wide variety of circumstances and will help you get your process back into control.

Verifying the Effectiveness of Corrective Actions

Whenever a process is adjusted, a note or event marker should be put on the chart. Most SPC chart software allows for either notes or some indicator of process changes on the charts. Charts made by hand can easily have notes written on them. In this way the charts

TABLE 4-3
Chronic Average-Related Symptoms and Causes

SYMPTOMS	CAUSES
Average frequently not in center of control limits	Process has been changed since the control limits were last calculated Limits were calculated on insufficient data Control limits were incorrectly calculated
Process stays within control limits but frequent changes in average occur	A process parameter that affects the average is frequently changing (e.g., material lot change or tool replacement) Frequent changes in measurement method or personnel
Average gradually changes	Something is wearing out during the process run Gradual environmental change is occurring Process parameter is gradually changing

TABLE 4-4
Effective Corrective Actions

CAUSES	CORRECTIVE ACTIONS
Inadequate training	Group trainees by job function and math ability. Then develop training program in house customized by job function using real data and real examples from in house Have refresher courses at fixed intervals like 6 or 12 months

EFFECTIVE CORRECTIVE ACTIONS 53

CAUSES	CORRECTIVE ACTIONS
	Have trainees take an SPC training test with awards and public recognition for those who pass
Insufficient time for SPC	Use SPC software Incorporate time for SPC when planning production activities and include in calculation of production time standards Set up SPC work area right at each production workstation—not in a separate room and not in a centralized SPC area
Lack of chart visibility or response to charts	Person plotting the SPC chart must be the same person responsible for adjusting the process Post charts prominently at eye level at each workstation. If software is used, consider additional monitors at eye level in heavy traffic areas
Insufficient data for calculations	Standardize minimum data amounts in written policies and procedures. Use minimum 25 subgroups
Incorrect calculations	Improve training Issue employees statistical calculators and train them to use them Use SPC software
Method of choosing what to do SPC on is incorrect	Have engineer do failure mode effects analysis (FMEA) and design of experiments (DOE), and do SPC only on most important parameters
PC practices not standardized	Document procedure simply and directly
Distribution not normal	Make histogram on recent data and interpret it per chapter 2

become a more valuable historical record of the process run and become a means by which the effectiveness of corrective actions can be verified and studied.

SPC measurements must always be made following any ongoing process adjustment. If the adjustment is one that will have an immediate effect, then take an additional unscheduled subgroup so you can verify that it had the desired effect. If the process adjustment takes a while to show any effect, then take several unscheduled subgroups during that time. This is so you can pinpoint the time at which the adjustment takes effect. Then in the future you need only take one extra subgroup at the appropriate time to verify the adjustment has worked. Once the effect of the adjustment shows on the control chart, track how long before you make that particular adjustment again. Over a period of time, this can give you a schedule of when to make the adjustment *before* it becomes necessary. This gives you a preventive process maintenance schedule that you can apply to actually prevent the process from going out of control in the first place.

If the change in average you are attempting to produce is quite small, then a statistical test of significant difference may be required to verify the average has indeed changed. Consult with a statistician or statistical reference book for further information on this.

Always check the control chart to verify that the process is back in control. Remember you must have at least 25 consecutive subgroups within the control limits after the corrective action in order to verify that the process is still stable. Your corrective action is not complete if it does not cause the process to remain stable.

If the problem was a lack of normality in the histogram, you will also need to construct a new histogram using at least 40 to 50 samples to check for normality. Always check the CpK after stability has been demonstrated following a corrective action, to be sure the process is still capable.

Conclusion

Thoroughly training the appropriate people and performing SPC correctly are the two most important steps toward preventing new problems from developing. Whenever possible, perform SPC on process parameters rather than product characteristics. Make the processes capable and stable before you start control charting. Training should emphasize chart interpretation and responding to out-of-control situations rather than chart construction or statistical theory. The person who actually controls the process is the one who should plot and respond to the charts. Choose the right kind of chart and apply the appropriate rules for interpretation and response.

All SPC systems must be monitored at both the system level and the chart level. System-level evaluations should occur a couple of times a year. Chart-level evaluations should be at least monthly. At the first sign of a problem, troubleshoot the SPC system right away.

GLOSSARY

ANOVA—(ANalysis Of VAriance) A statistical method that analyzes and compares data from different groups.

assignable cause—Any cause of variation other than normal random variation.

attribute—A kind of SPC where the data plotted on the chart is counted rather than measured. *p, np, c,* and *u* charts are attribute charts.

capability index—A number that relates the amount of variation in data to its specification. Higher index numbers indicate less variation.

central limit theorem—A theorem of statistics that says when a group of averages are calculated from a group of samples that are all taken from the same parent population, the averages will be normally distributed.

chart evaluation—A type of SPC audit that verifies the SPC chart is properly constructed, used, interpreted, and that it recieves adequate responses when necessary.

combed distribution—A histogram shape that has alternating high and low values that make the appearance of teeth on a hair comb.

design of experiments—A standard statistical method of planning and performing experiments with more than one variable and each variable set at more than one level. Usually this is done as a factorial or partial factorial experiment.

exponential distribution—A distribution or histogram shape resulting from a singular cause of change and having a characteristic exponential graph shape with the highest point at one end and the lowest at the other.

gage R&R—The combined effect of variation from gage repeatability and reproducibility.

gage repeatability —The variation in measurements observed when one person measures the same thing, in the same way, using the same gage. The total spread of the distribution of the same measurement repeated over and over on the same device by the same person is the repeatability of the measurement on that device.

gage reproducibility—The maximum difference in the averages of each person's data, when each person measures the same thing, in the same way, using the same gage. The difference between the averages of repeated measurements of the same thing, made by different people, using the same gage, is the gage reproducibility for that gage on that measurement.

kurtosis—The relationship between the steepness of the sides of a bell-shaped distribution curve and the length of its tails.

multimodal distribution—A histogram or distribution curve shape that has more than one distinct high point as if normal bell curves are overlapping.

multivariate distribution—A histogram or distribution curve shape that has no distinct high point or tails and is somewhat flattened in appearance. It results from the influence of many different variables.

nonrandom pattern—Any pattern that has an assignable cause or repeats over and over again.

Pareto chart—A bar graph of frequency of occurrence sorted from highest to lowest.

part parameter—A physical characteristic of the part produced by the process. Examples are length, weight, hole diameter, etc.

process parameter—A characteristic that when adjusted influences the process. Typical process parameters are temperature, pressure, and speed.

RMS deviation—Standard deviation calculated by the RMS method rather than estimated by a factor from a table.

system evaluation—A type of SPC audit that examines the SPC system as a whole. It checks the planning and design of the SPC system itself. This includes among other things, training, process capability studies, and procedures (not only that procedures are applied correctly, but that the procedures have merit of their own).

variables—A kind of SPC where the data plotted on charts is measured rather than counted. X-bar and R charts are examples of variables SPC.

Weibull distribution—A distribution calculated by a scale parameter known as alpha (α), a shape parameter known as beta (β), and a location parameter known as gamma (γ). These values, calculated from separate equations, are placed into the Weibull distribution equation to create the Weibull distribution, which is much used in reliability statistics.

RECOMMENDED READING

For information on how to perform SPC

Amsden, Robert & Davida and Butler, Howard; *SPC Simplified, 2nd Ed.;* Quality Resources, New York NY, 1998.

Automotive Industry Action Group; *Fundamental Statistical Process Control;* Southfield MI, 1991.

Shewhart, Walter A.; *Economic Control of Quality of Manufactured Product;* van Nostrand, 1931; republished by American Society for Quality Control, Milwaukee WI, 1980.

Western Electric Co., Inc., *Statistical Quality Control Handbook;* Western Electric Company Indianapolis IN, 1956.

For relating SPC to other statistical methods

Automotive Industry Action Group; *Measurement System Analysis;* Southfield MI, 1990.

Enrick, Norbert Lloyd; *Quality Control and Reliability;* Industrial Press, Inc., New York NY, 1977.

Juran, J. M. and Gryna, Frank M., Jr.; *Quality Planning and Analysis;* McGraw-Hill Book Company, New York NY, 1980.

For information on applying statistics to find root causes

Juran, J. M. and Gryna, Frank M., Jr.; *Quality Planning and Analysis;* McGraw-Hill Book Company, New York NY, 1980.

Snedecor, George W. & Cochran, William G.; *Statistical Methods;* Iowa State University Press, Ames IA, 1980.

Squeglia, N. L.; *S.P.S. - Part One Experimental Designs Made Simple;* N.L. Squeglia, Orange CT, 1988.

INDEX

Actual standard deviation (RMS deviation), 15, 19-20
ANOVA (analysis of variance), 36, 50, 51
Audits, quality, 7-8
Average
   actions for chronic issues with, 50-51
   calculation of, 32
   CpK (process capability with respect to specification closest to process average), 8-10, 14, 15, 16, 17, 19, 35, 46, 51, 55
   long-term change in, 31-32
   symptoms and causes of issues with, 52

Binomial distribution, 26-27

Capability indexes, 13-20
   Cp (process capability), 13, 15, 16, 17, 46
   CpK (process capability with respect to specification closest to process average), 8-10, 14, 15, 16, 17, 19, 35, 46, 51, 55
   CpL (process capability with respect to lower specification), 14, 15, 17, 19
   CpM (process capability with respect to nominal), 14-15
   CpU (process capability with respect to upper specification), 14, 15, 17, 19
c charts, 30, 31-32
Central limit theorem, 42
Chart evaluations, 10-11
Combed distributions, 24, 34
Control charts
   chart level evaluations and, 10-11
   excessive number of, 48
   impact of implementing, 6-7
   lack of control in, 5
   lack of response to, 48
   lack of visibility of, 53

Control charts (*cont.*)
  non-normal distributions and, 20-27
  policy and procedure differences between departments, 48
  in prevention of defects, 3
  principles for using, 29-31
  purpose of, 34-35
  rules for interpreting, 30
  sample checklist, 9, 10
  variation in, 2, 6-7
Controllability, 2
Control limits
  calculation of, 31, 32, 35, 36-37
  control limit factors, 19-20
  data points outside, 31-35
  incorrect, 48
  math errors in, 48
Coordinate measuring machine (CMM), 39
Corrective actions, 43-55
  for chronic average issues, 50-51
  for excessive variation, 49-50
  failure of, 43-46
  for SPC issues, 46-49
  verifying effectiveness of, 51-55
Costs
  of SPC, 5, 6
Cp (process capability), 13, 15, 16, 17, 46
CpK (process capability with respect to specification closest to process average), 8-10, 14, 15, 16, 17, 19, 35, 46, 51, 55
CpL (process capability with respect to lower specification), 14, 15, 17, 19
CpM (process capability with respect to nominal), 14-15
CpU (process capability with respect to upper specification), 14, 15, 17, 19
Cut distributions, 24-25
Cyclic patterns, 38-39

Defect rates
  defect prevention and, 3
  of defects versus defective parts, 32-33
  variability of, 5, 6
Design of experiments (DOE), 18, 36, 50, 51, 53

Edge-peaked distributions, 24-25
Employee complaints, 5, 6
Engineers
   training for, 6
Errors
   checking SPC charts for, 10
   mathematical, 32
   in practice of SPC, 32
   in SPC calculations, 5, 6, 53
   SPC software and, 32
Expectations
   customer, 2
   for statistical process control, 1
Exponential distributions, 25-27

Failure mode effects analysis (FMEA), 53

Gage reproducibility and repeatability (R&R), 17
Goals
   process, 2
   product, 2-3

Hypergeometric distributions, 26-27

Internal quality audits, 7, 8
ISO 9001, 7

Kurtosis, 10, 19, 21-22
   compensating for, 21-22
   defined, 21
   leptokurtosis, 22
   platykurtosis, 22
   positive and negative, 21, 32

Leptokurtosis, 22

Maintenance, preventive, 38
Managers
   training for, 6
Median
   defined, 19
   in normalization of skewed distributions, 19, 20

Mode
   defined, 19, 20
   in normalization of skewed distributions, 19, 20
MR charts, 32
Multimodal distributions, 22-23
Multivariate distributions, 23-24

Negative kurtosis, 21
Negative skewness, 22
Noncompliances, 5, 6
Non-normal distributions, 20-27
   combed, 24, 34
   corrective actions for, 53
   cut, 24-25
   edge-peaked, 24-25
   eliminating assignable causes of variation, 19, 20-21, 29, 30, 37, 45
   exponential, 25-27
   kurtosis in, 10, 19, 21-22, 32
   multimodal, 22-23
   multivariate, 23-24
   skewness in, 10, 19, 20, 21-22, 25-26, 30-31, 33
   Weibull, 25-27
Nonrandom patterns, 37-42
   cyclic, 38-39
   run sum violations, 39-42
np charts, 31-32

Omissions
   checking SPC charts for, 10
Over-control, 34

Parameters
   process, 6
Pareto charts, 10
p charts, 31-32
Platykurtosis, 22
Poisson distributions, 26-27, 30
Positive kurtosis, 21, 32
Positive skewness, 22, 25-26
Pp, 15

PpK, 15
Predictability, 2
Preventive maintenance, 38
Process adjustments, 5, 6, 7
   process changes versus, 37
   SPC measurements following, 54
Process capability, 5, 6, 13-27
   measuring, 13-15
      Cp (process capability), 13, 15, 16, 17, 46
      CpK (process capability with respect to specification closest to process average), 8-10, 14, 15, 16, 17, 19, 35, 46, 51, 55
      CpL (process capability with respect to lower specification), 14, 15, 17, 19
      CpM (process capability with respect to nominal), 14-15
      CpU (process capability with respect to upper specification), 14, 15, 17, 19
      Pp and PpK, 15
   modification for non-normal distributions, 19, 20-27
   process centering and, 16-17
   process variation and, 17-18, 49-50
   recalculating, 35, 36-37
Process capability studies, 34
Process centering, 16-17
Process changes
   process adjustments versus, 37
Process goals, 2
Process parameters, 6
Process variation, 17-18
   corrective actions for excessive, 49-50
   symptoms and causes of excessive, 50
Product goals, 2-3
Productivity, 3
   variability of, 5, 6

QS-9000, 7
Quality audits, 7-8

Range, 32
R charts, 20, 30, 32, 33, 42
Recertification, 10

Regression analysis, 36, 50
Resolution, 24
RMS deviation (actual standard deviation), 15, 19-20
Rounding off, 24, 34
Runs, chronic, 35-37
Run sum violations, 39-42

Sampling frequency, 30-31, 32
S charts, 32
Skewness, 10, 32-33
   compensating for, 21-22
   defined, 21
   normalization of, 19
   positive and negative, 22, 25-26
   subgroup size and, 30-31, 33
SPC software, 20, 32, 33, 53
Statistical process control (SPC)
   causes of system problems, 48
   corrective actions for, 46-49
   expectations for, 1
   most effective use of, 6
   normality and, 18-27
   preliminary work in, 18-19
   process goals in, 2
   product goals in, 2-3
   troubleshooting in. *See* Troubleshooting
Subgroup size, 30-31, 33
Supervisors
   training for, 6
System evaluations, 8-10
   See also Process capability

Timing
   of process adjustments, 7
   of quality audits, 7-8
Training, 6
   as corrective action, 52-53
   identifying need for further, 10
   measuring effectiveness of, 8-9, 10
   in statistical process control (SPC), 47-48
   verifying effectiveness of, 8-9

Trends, chronic, 35-37
Troubleshooting, 5-11
    chart evaluations and, 10-11
    symptoms of need for, 5-8, 31-42
    system evaluations and, 8-10

u charts, 30
Uncontrolled processes
    chronic runs or trends, 35-37
    data points outside control limits, 31-35
    nonrandom pattern of, 37-42
    symptoms of, 5-8, 31-42

Variables
    multivariate distributions and, 23-24
Variation
    control chart, 2, 6-7
    corrective action for excessive, 49-50
    eliminating assignable causes of, 19, 20-21, 29, 30, 37, 45
    process, 17-18

Weibull distributions, 25-27
Western Electric Rules, 30

X-bar charts, 20, 30, 32, 33, 42
X charts, 32

Also available from Quality Resources ...

SPC Simplified: Practical Steps to Quality, Second Edition
Robert T. Amsden, Howard E. Butler, and Davida M. Amsden
304 pp., 1998, Item No. 763403, paperback

Everyone's Problem-Solving Handbook: Step-by-Step Solutions for Quality Improvement
Michael R. Kelly
165 pp., 1992, Item No. 916528, spiral

Safety Management and ISO 9000/QS-9000: A Guide to Alignment and Integration
Robert J. Kozak and George Krafcisin
184 pp., 1997, Item No. 763179, paperback

Implementing the TE Supplement to QS-9000: The Tooling and Equipment Supplier's Handbook
D.H. Stamatis
368 pp., 1998, Item No. 763537, hardcover

The ISO 9000 Book: A Global Competitor's Guide to Compliance and Certification (Second Edition)
John T. Rabbitt and Peter A. Bergh
211 pp., 1994, Item No. 76258X, hardcover

The Miniguide to ISO 9000
John T. Rabbitt and Peter A. Bergh
59 pp., 1995, Item No. 763020, paperback

The QS-9000 Book: The Fast Track to Compliance
John T. Rabbitt and Peter A. Bergh
240 pp., 1998, Item No. 763349, hardcover

The QS-9000 Miniguide
John T. Rabbitt and Peter A. Bergh
88 pp., 1997, Item No. 763233, paperback

The Basics of FMEA
Robin E. McDermott, Michael R. Beauregard, and Raymond J. Mikulak
76 pp., 1996, Item No. 763209, paperback

The Root Cause Analysis Handbook: A Simplified Approach to Identifying, Correcting, and Reporting Workplace Errors
Max Ammerman
144 pp., 1998, Item No. 763268, paperback

For additional information on any of the above titles or for our complete catalog, call 1-800-247-8519 or 212-979-8600.

Visit us at www.qualityresources.com
E-mail: info@qualityresources.com

Quality Resources, 902 Broadway, New York, NY 10010